Monique Heurleur

D0731402

In this stimulating book Dolf Rieser draws attention to
the attributes shared by two such apparently divergent
disciplines as art and science, in particular indicating
those areas in science where art plays a part. His text
ranges from the study of micro-organisms and their
structures, to visual perception and artistic vision, as
well as a consideration of the use of art therapy in the
treatment of psychotics. An important chapter on art
forms in nature embraces the study of such fundamental
patterns as the electron-wave pattern, structural plant
forms and their functional patterns of growth and
rhythm. Rieser reveals those features of outstanding
beauty, harmony and balance to be found in many basic
patterns of natural growth.

Dolf Rieser is an artist who trained as a biologist. His
engravings, which have been widely exhibited in
Europe and the USA, reveal something of his preoccupation with biological and cellular structures.
Born in South Africa, he gained his doctorate in
science from Lausanne University before moving to
Paris to study art.

Art and science will be of particular value to artists
and designers. It will also be of absorbing interest to
all those involved with teaching and will stimulate
every reader to reconsider the established boundaries of
artistic and scientific activity.

Art and science

Dolf Rieser

modes of thinking
visual perception and artistic vision
art forms in nature
art and the unconscious mind

Studio Vista: London
Van Nostrand Reinhold Company: New York

Cover
The Spiral of Life, intaglio etching by Dolf Rieser, superimposed on a photograph of a cellular structure

A Studio Vista/Van Nostrand Reinhold Paperback
Edited by John Lewis
© Dolf Rieser 1972
Published in London by Studio Vista,
Blue Star House, Highgate Hill, London N19
and in New York by Van Nostrand Reinhold Publishing Company,
a Division of Litton Educational Publishing, Inc.,
450 West 33 Street, New York, NY 10001
Library of Congress Catalog Card Number: 79–151409
Set in 11 on 14 Univers
Printed and bound in the Netherlands
by Drukkerij Reclame N.V., Rotterdam
ISBN 0289 70065 5 (paperback)
       0289 70066 3 (hardback)

Contents

Acknowledgements
I am indebted to Dr Eliot Slater for advice given
regarding the functioning of the human brain. I also
wish to acknowledge the help received from the books
written by Professor Richard Gregory (University of
Bristol) and by Peter McKellan (University of
Sheffield). I am grateful to Mr Adrian Loveless for
taking photographs of specimens in may collection and
I am also grateful to Mr Martin Rieser for drawing the
diagrams and Mr Sven Rindl for devising a diagram of
a high-block structure.

I am indebted to the following for permission to
reproduce illustrations: Victoria and Albert Museum
(Mantegna, Palmer, Rembrandt); Tate Gallery (Blake,
Klee, Miro, Picasso); British Museum (Cylinder Seal);
Science Museum (for photographing model of the DNA
structure); Messrs Lumley Cazalet Ltd (Miro); Studio
Vista; Faber and Faber; Penguin Press; Phaidon Press;
and *Scientific American*.

1. Introduction: modes of thinking

Art and science are generally considered totally separate disciplines. The aim of this book is to draw attention to some of the qualities they share. It is not a comprehensive survey of either discipline but is intended to stimulate new ways of regarding the relationship and status of art and science.

The findings and interpretation of modern psychology illustrate one aspect of the pronounced influence of science on certain art expressions, in particular on the Surrealists but also on individual writers such as James Joyce and D. H. Lawrence. Of equal importance has been the impact of modern psychology on scientific thinking, where a widening outlook has led to a deeper comprehension of conceptions of physics. Logical explanations had hitherto formed the basis for all theories concerning natural phenomena. The ideas introduced by psychology aimed to find connections in nature rather than to look for causal laws. The eminent nuclear physicist, Dr W. Pauli, for instance, began to study the role of symbolism in the field of scientific concepts. He believed that investigations of outer objects should run parallel with psychological investigations regarding the origin of scientific concepts.

Teilhard de Chardin said: 'The spirit of research and conquest is the paramount soul of evolution.' The spirit of research permeates all those who have ever dedicated their intelligence and life to art or science.

The human mind is activated by a permanent sense of interest and curiosity concerning all existing phenomena. This all-prevailing sense led to the greatest discoveries and to the most important human expressions.

As disciplines of the human mind science and art are analogous as well as prominently different. From the earliest ages, art has been a profound manifestation of living people. All civilizations have expressed ideas in their own respective artistic idiom, and have consciously evaluated such creative expressions. Science, on the other hand, tries to elucidate natural processes which follow fundamental laws. Each specific process is directed by such laws which science tries to examine by varying means adapted to the particular science. Science is, therefore, the investigation and establishment of general laws concerning the behaviour of the world we inhabit. Scientific truths are expressed in an abstract, mathematical language capable of describing relevant discoveries and experiences. It is not enough to evolve theories without being able to prove them by logical deduction and by practical observation and research; all theories have to be modified with the advent of new scientific discoveries.

It is impossible to enlarge here on the relative value of the various scientific methods used, whether they be deductive or inductive, or on the philisophy of science in general. At present, however, the so-called 'hypothetico-deductive method' appears to be the most promising method of approach. Here, theory, observation and imagination cooperate to arrive at results. It is now evident that the imaginative element plays a major role in elucidating natural phenomena and establishing general laws. So that, at least in the initial stages, there seems to exist a very close and almost parallel development of scientific or artistic ideas. For the scientist, the idea or inspiration may appear very suddenly, almost out of the blue, suggesting, therefore, that the ideas themselves can derive from the deepest parts of the human mind. The idea, however, could not materialize without a number of separate facts being connected in the mind by an intuitive process. The scientist will then substantiate his hypothesis by tests and experiments. To put it colloquially, the 'hunch' comes first.

Although artistic ideas evolve along similar lines in the initial stages, their development seems to differ considerably from scientific conceptions, mainly because subjective elements play a more important role than objective and verifiable elements. Yet artists too, for instance Klee and Kandinsky, through numerous manifestoes and writings, have attempted to describe and solve the problems of perception and expression.

Imagination

As well as reasoned and logical thinking, imagination plays a major role in any creative activity. The 'creative imagination' has been investigated a number of times, particularly by psychologists. The terms A-thinking and R-thinking distinguish between two different kinds of thinking. The category of A-thinking (so named by the psychologist Bleuler) refers to that kind of thinking which includes visions, hallucinations, dreams, nightmares and so on; images in fact which also appear under the influence of psychotic products.

R-thinking, on the other hand, includes the critical and logical types of thinking. P. McKellar suggested that the most useful products of thought would develop by an interplay of A- and R-thinking processes. But this theory was considered inadequate in some cases since certain logical inferences begin from false assumptions. In some forms of thinking, encountered in psychosis for instance, this is certainly the case. Nevertheless, it seems that an interplay of these two kinds of thinking is responsible for both scientific and artistic productions. An example of this can be found in the work of the chemist, Kekulé, who developed the important theory of the Benzene Ring. The 'idea' came to him on two occasions when he was half asleep and saw the image, as he put it, 'as dancing atoms whirling in a ring, the larger ones forming a chain dragging the smaller ones'. This kind of imagination is now termed 'hypnagogic' (i.e. images occurring during half-sleep). Kekulé persisted in critically testing his dream by logical intelligence when wide awake. This of course entailed so-called R thinking.

A prevalence of one particular type of thinking process possibly affects personality. In artistic professions, for instance, the decision to become either a designer or an artist may be influenced by R- or A-thinking respectively.

Visionary imaginations were surely possessed by such great artists as Hieronimus Bosch and William Blake. In modern psychological terms, Blake must also have

1 *Elohim creating Adam* by William Blake
By courtesy of the Tate Gallery, London

had an introverted, intuitive personality. If people such as Blake happen to be creative artists, they produce, as Blake did, a visionary world all of their own. Blake was a mystical dreamer able to create his own visions in visible shapes.

Experiments carried out with hallucinogenic drugs have led to many interesting discoveries regarding the imagination in its A-form. Some such drugs may have formed the famous 'potions' of witches in medieval times. The pre-Mexican Indians used Peyote (Mescalin) to induce hallucinations during their religious ceremonies. It would now appear that drugs produce a number of changed mental states and phenomena which are partly manifested as visual imaginations, hallucinations or thought disturbances. The drug, Mescalin, produces psychotic states resembling schizophrenia and this, in turn, may throw some light on true psychotic states.

The visions and hallucinations brought about by such drugs as LSD and Mescalin, do so because they intensify the senses. Such visions can often be very beautiful, assuming vivid colours and extraordinary shapes. Images produced in this manner can repeat themselves in recurrent, identical shapes such as spirals, arabesques or webs. One investigator, H. Kluver, called them 'form constants'. There seems to exist here an analogy with recurring shapes originating in the inner eye, such as the phosphenes described in Chapter 2. The drug LSD 25 can induce psychic disintegration. Experiments have shown that the drawings depicting the same objects and executed by the same artist became progressively more non-figurative and abstract under the influence of this drug, as the conscious control diminished and was slowly replaced by the 'unconscious'.

Serious, psychological investigations have been carried out in these directions. It has been found that imagination covers a wide field of different activities and states of mind, which, in turn, form the sources of many creative ideas such as, for instance, the dream imagery already mentioned. Examples of similar imaginary images can be found in the work of some Surrealist painters such as Salvador Dali and Max Ernst.

Mental images
Mental images should not be confused with after-images (see Chapter 2). It is assumed that the ability to perceive imaginative pictures in the mind is based on visual experiences: 'a previous perceptual experience

can be reproduced in the absence of the original sensory stimulation'. Visual memory can be advantageous in some professions. Artists rely on this ability which is further developed by constant use. However, there exist a number of different processes of remembering images which in turn produce different kinds of mental pictures. It has also been found that there are different possibilities regarding 'day dreaming'.

One line of scientific research is investigating the differences which exist between images experienced during sleep and images experienced during waking hours. One method of investigation is based on the study of eye movements which take place during dreaming. These are called REM (rapid eye movement). New techniques allow for the study of such movements during sleep and also during waking hours. Electrical recording machinery, used by American investigators, showed that eye movements increased during periods of logical thinking in contrast to the thinking taking place during the 'day dreaming'.

Creative thinking
Imagination leading to creativity is closely connected with the representation of inner symbols in visible form. Such symbols called 'archetypes' by psychologists play

a considerable role in the world of scientific concepts and are also to be found in numerous art expressions. Archetypes represent unconscious images common to all men.

A number of examples show how particular problems have been solved by the interaction of conscious and less tangible sources of inspiration. Henri Poincaré, the great mathematician, described how he arrived at a solution to a mathematical problem after a long period of fruitless research. During a restless night an image of changing atoms appeared to him. Next morning, almost in a flash, he found the solution which terminated his investigations. Poincaré describes the progressive steps of the creative process:

(a) the initial period of conscious work dealing with the problem;
(b) a period during which the unconscious mind seems to be active when an appropriate hypothesis strikes the thinker with its aesthetic properties much as a good work of art does. Aesthetic sensibility is the clue to the soundness of a hypothesis;
(c) proof of the hypothesis has next to be worked out.

Similar experiments and experiences were reported by Ghiselin. Experiments undertaken by Patrick (1936) involved giving creative persons such as painters or poets subject matter to work on and asking them to report, at the same time, on their activities. It was found that creative thinking follows certain lines in definite stages. The personalities of the artists involved seemed to play a major role. Poincaré summarized it in the following way: 'A period of preliminary activity precedes all fruitful, unconscious work.'

Beauty in science
Scientists often describe scientific theories as very beautiful. Some mention this as their most outstanding reward. This aesthetic sense in some cases serves as a guiding line to their problems, by the arrangement of facts into an ordered harmony, where formerly was disharmony and disorder. For example, the harmony of numbers and the geometric beauty of forms have long been recognized. Even Pythagoras was aware of musical harmonies based on the length of strings standing in sequential relationship.

Einstein said: 'Beauty is the first test. There is no place in the world for ugly mathematics.' Poincaré also mentioned that scientists are motivated by pleasure in their studies, experiencing this pleasure because nature is beautiful. He was referring to a more 'intimate beauty, which comes from the harmonious order of parts, which a pure intelligence can grasp'.

One of the scientists closely linked with research into quantum mechanics, P. A. M. Dirac, said that beauty was a safe guide in the search for truth and, further, that it has always been found that highly successful rules are also beautiful. A further example mentioned by Dirac refers to Schroedinger's wave mechanics (see Chapter 2) as a beautiful way of gathering the essential data for the building of the quantum theory. At first these did not coincide with the facts, but later it was discovered this was due to some new, unknown factors.

2. Visual perception and artistic vision

Apart from the sense of hearing, the eyes are the most important sense organs. Seeing is a complicated process dependent on the intimate collaboration of eyes and brain. Any visual information received is registered, sifted and classified in the visual area of the brain. Furthermore, the eyes are the means by which images conceived by the human mind can be materialized in tangible form. 'Artistic vision' is the creative process which leads to visual images.

Perception depends on numerous factors, partly psychological and partly based on cultural traditions. In artistic perception a guiding principle can further complicate the issue. Some examples are given to indicate how the appearance and expression of creative images was altered by outer circumstances and conditions.

Recent scientific investigations have progressively increased the understanding of how ideas are materialized. In an attempt to analyse the motivations behind artistic expression, the following questions arise: first, what interaction exists between the ideas, their development and manifestation in visible form? Secondly, how did the eyes and a particular way of perception influence the expression of such artistic ideas?

The eyes

The eyes receive and register radiant energy or light-waves and allow us to see. Light is propagated by electro-magnetic waves, but the visible spectrum to which the eye is sensitive forms only a very small part of the entire scale of electro-magnetic waves. These range from the smallest imaginable to mile-long wavelengths. The visible spectrum ranges from 400 to 700 millimicrons (or from 400 to 700 billionths of a metre). Only these wavelengths can be registered by the human eye (see fig. 2). The process of seeing, as such, can be divided into two closely linked parts. First, seeing involves a purely physical stimulation which is produced by light rays entering the eye and acting on the retina. The second part involves the visual area of the brain where light impulses forming images are coordinated and interpreted. All occurrences seen by the action of the eyes are conveyed to the brain, but the radiation impulses are translated into nerve impulses in the optic nerve. In this way, they reach the respective brain areas.

The basic rhythm of human life and activity depends very much on this system which communicates all visible facts to the brain. Conversely, vision is greatly affected by psychological factors.

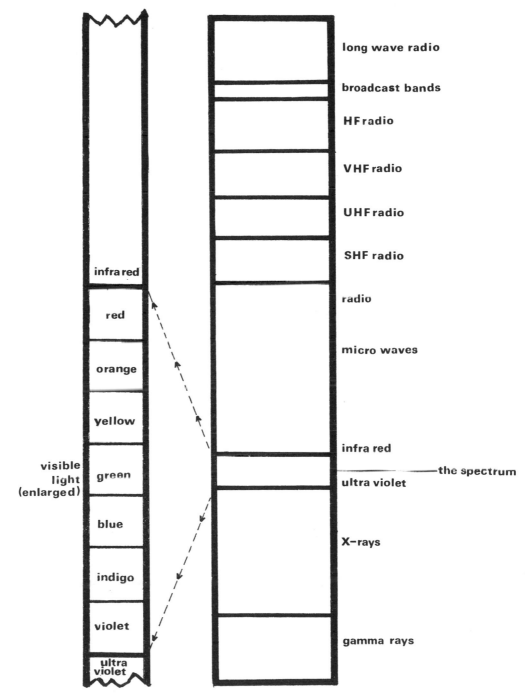

infra red

red

orange

yellow

visible
light
(enlarged)

green

blue

indigo

violet

ultra
violet

long wave radio

broadcast bands

HF radio

VHF radio

UHF radio

SHF radio

radio

micro waves

infra red

————the spectrum

ultra violet

X-rays

gamma rays

Eye structure

The light rays entering the eye act on the retina in the form of a photo-chemical stimulus. Two kinds of retinal receptor cells, the so-called 'rods' and 'cones', are mainly concerned with this process. The rod cells contain the pigment 'visual purple', while some other retinal cells contain a brown pigment. Lightwaves entering the retina have the effect of moving the brown pigment granules, of shortening the rods and bleaching the brown pigment. Furthermore, the visual purple is bleached and transformed to visual yellow by the light. The rods, having thus been chemically stimulated, pass on the stimuli to the nerve fibres of the optic nerve.

The second important light receptor cells of the retina consist of the cones. These cells deal mainly with the reception of high-intensity stimuli or, in other words, with 'daylight' vision, sensitive and responsible for the recording of colours (cone vision).

The rods dealing with less intense light frequencies are concerned with 'twilight' vision (rod vision). Radiant light energy is received by the millions of light-sensitive cells of the retina and is then converted into nerve impulses. In connection with 'twilight vision', it is interesting to note that the human eye, when it adapts to the dark, becomes insensitive to the colour red. Red is therefore seen as black even in half light. This negative reaction to the red pigment can have considerable practical consequences, in particular where traffic is concerned.

The visual receptor cells in the retina are densely packed. The light rays which enter the eyes have to pass through a layer of neurons, which form a network with the photo-receptors and also with each other. All these nerve cells end in the optic nerve leading to the brain. The receptor cells, in responding to the light stimulus, set up physio-chemical conditions which produce and release impulses which then travel along the nerve fibres in the form of electrical impulses. This is made possible through the structure of the fibres which contain negative potentials inside and positive potentials outside.

Colour perception
Only a few facts concerning colour perception are discussed here. Leonardo da Vinci, Descartes, Newton, Goethe, Van Helmholtz, Thomas Young and others studied and established various colour theories. The most reliable theory, even today, was first developed by Thomas Young in 1801 and then expanded by von

Helmholtz. The problem of colour perception is summarized by Professor R. L. Gregory in the following way:

Newton had already pointed out that white light is composed of all colours of the spectrum and that each colour corresponds to a given frequency. The eye has to produce a different neural response to different wave frequencies. However, light frequencies are extremely high in the visible spectrum and are, therefore, higher than those transmissible by nerve impulses.*

Certain receptor cells respond to colours, in particular to the 'primary' colours which Young considered to be red, green and blue. He found that he could produce any colour of the visible spectrum by a mixture of three lights, set to appropriate intensities. Any spectral hue can be produced by adjusting relative light intensities. It follows that there must be three kinds of colour-sensitive receptor cells ('the cones') which respond respectively to red, green and blue. All colours are seen as a mixture of nerve signals from the three systems. Only lightwaves produced by a prism, or those retained after passing through a coloured filter, are considered here. The yellow and blue mixed by painters to produce

* R. C. Gregory *Eye and Brain*. London: Weidenfeld and Nicholson 1966.

green is a question of colour pigments. A painter mixes the total colour spectrum minus the colours absorbed by the pigments.

An important feature which the Neo-Impressionist painter Seurat understood, and which has been applied by many artists, is that the relative value of the perception of colours is much influenced by their setting. One colour pigment can alter the perception of surrounding pigments. Complementary colours intensify each other. A blue pigment, for example, appears more intensely blue next to its complementary yellow. A coloured disc situated in a purple field looks quite different when compared with the same disc situated in a yellow field. Both colour appreciation and judgment depend not only on the surrounding colours but also on the type of illumination.

Phosphenes
The appearance of 'phosphenes', colloquially called 'seeing stars', is well known to everyone. On entering a completely darkened room, colour spots start to appear in the eye, once the eyes have become accustomed to the darkness. The same happens when artificial pressure is put on the eyes. In each case, colour patterns and shapes appear which do not enter the eye

in the normal way, but are produced within the eye and brain. With the help of phosphenes it is possible to study the functional organization of the brain, and recent research by Penfield (Montreal Neurological Institute) established that phosphenes originate all along the visual pathway and that it is possible to stimulate visual areas in the brain to produce such phosphenes. Stimulations of this kind produce images and visual experiences of the past. Other investigations found that patients who had been blind for a long time began to see phosphenes after similar treatment. It was not possible, however, to achieve such results with persons who had been blind since birth. It appears then that electrically produced phosphenes can be grouped into definite pattern categories. The shapes that appear are nearly all of geometric design, frequently resembling the scribbles produced by young children.

As a phenomenon, only recently fully investigated it is thought that phosphenes may have had some unsus-pected influence on the origin of certain artistic images of the past. Some geometric forms in art produced in different countries, and at different historical periods, may have a common origin in phosphene shapes. It is certainly significant that identical geometric shapes such as circular patterns, regarded as archetypal symbols, appeared for instance in prehistoric paintings and, very much later, in clay stamp patterns produced in Mexico.

Disturbing patterns

Certain patterns and designs are extremely disturbing to the human eye. Closely spaced parallel lines can irritate the eyes and the same happens with so-called 'ray' figures. D. M. McKay has studied this problem and suggests that the visual system is upset by the redundancy of similar designs. He assumes that, normally, the redundancy of the object is used by the visual system to save itself the work involved if the information has to be analysed. In such cases, the redundant figures are so extreme that the visual system is upset by them. The whole problem is not yet completely solved. This may depend to a certain extent on eye movements. It does seem that the visual system is much upset by similar patterns. This may be the reason why certain 'Op-art' designs produced today have such disturbing effects, which some artists, such as Bridget Riley and Victor Vasarely, have used quite deliberately.

Optical illusions

The mechanisms of optical illusions have long been investigated. A simple illustration can be given by the Müller-Lyre illusion. The central lines are exactly identical in length but the line bearing the two normal arrows appears shorter.

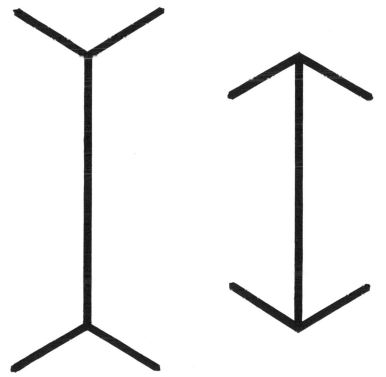

5   The 'Müller-Lyer illusion'. The shafts of both arrows are identical in length

Another well known example of illusion is illustrated by the so-called distortion room constructed by Adalbert Ames. Here, misleading items confuse the observer and his established mental images to such a degree that he cannot perceive correctly. Any things or persons in the room seem to be of different sizes, but at identical distances, although in reality they are at different distances. A most astounding figure can be seen in fig. 6, which shows a configuration of three hollow tubes. It would seem impossible to relate the centre pipe to any object in existence.

6   The three-pronged illusion

Perceptual phenomena
Stimulation of the eye can lead to after-images, such
as the dark patch which appears in front of the eyes
when a bright source of light has been looked at. Many
such experiments can be undertaken. After-images in
colour are known as positive when they appear in the
same colour they are seen in. Negative after-images, on
the other hand, appear in the relative complementary
colour. They occur when coloured planes have been
looked at for some time in bright light and then the gaze
is placed on to a white plane. Thus a red stimulus
would produce an after-image of its complementary
green.

The brain
The functioning of the visual area: converted nerve
impulses are conveyed by the optic nerve fibres from
the retina to the brain. The human brain consists of two
halves, the left and the right hemispheres, connected by
nerve tissues. When separated by surgical operation,
the two halves function independently as a complete
brain. Research by American and Italian scientists has
found that the right brain hemisphere is inferior to the
left in the command of language. In experiments on
monkeys it appears that split-brain animals can handle
more visual information than normal animals. Visual

stimuli in the left visual field could not be described by humans who had undergone split-brain operations, because of the disconnection between the left and right hemispheres and the separation of the speech centre which is in the left half of the brain.

The nerve fibres from the right half of each eye go to the right side of the brain, whereas those from the left half of each eye go to the left. It follows that they cross at one point. There are a number of essential nerve links along the pathway from the eye to the brain. Very little is known so far about the way in which the millions of brain cells interact, and which are the sensory paths which transmit impulses from the eye.

There is a control centre in the visual area of the brain cortex which coordinates the intake transmitted by the optic nerve. This centre seems, in turn, to be controlled by deeper lying control centres where information is stored. The complex structure of the brain neurons enables it to detect the numerous impulses coming in and to establish an overall picture. The brain seems to act as a complex computer system which sorts out and coordinates facts and also stores or eliminates all incoming information.

A comparatively large area of the brain is devoted to vision alone. The central area of the retina is densely covered with receptor cells which absorb the finest details. The area of the visual cortex of the brain is 10,000 times as large as the retinal area. This alone indicates the great amount of brain activity involved solely with vision.

Little is known as to how the visual region of the cortex coordinates impulses. According to Gregory, it might be assumed 'that retinal patterns are represented by coded combinations of cell activities'.

Learning how to see
Children learn how to see by acquiring the ability to recognize objects. Visual perception alone would not convey even the full physical meaning of unknown objects. It is possible to identify seeing with recording in the mind. Visual perception, as such, can and does alter according to acquired knowledge. Blind persons who regain their sight have to learn again how to see. A case studied by Gregory made it clear that information acquired by touch during blindness greatly helped the person to identify objects by sight. It was a different matter where distance was involved, which proved extremely difficult to assess. In the case of

persons whose two brain hemispheres had been separated by surgical operations it was observed how quickly the brain learned to see and adapt itself to new circumstances. The ability of the brain to absorb and adapt itself to new aspects of vision seems to be the basis of all visual learning and interpretation. Infants do not possess acquired knowledge to help them develop visual conceptions. So far, little is known of how they acquire the ability to recognize objects and learn how to perceive.

Artists extend their visual perception by constant and intensive use of the eyes. It does seem possible to develop the sense of depth and acquire a more intense awareness of the relationship between objects. A theory explaining how the eye sees depth and distance was recently developed at Berkeley University in California. Activities of single nerve cells in the visual area of the brain were recorded. It was discovered that the retina is connected to the brain cells in an inaccurate way. It is precisely this inaccuracy which would enable certain brain cells to react to objects from a range of vantage points and which stimulates stereoscopic vision.

7 *Adam and Eve* by William Blake
By courtesy of the Tate Gallery,
London

Artistic vision

Psychological research has made clearer the origins
which motivate all art. It would appear that creative,
artistic vision is most often 'an expression of a world
behind consciousness'. This would explain why it
seems impossible to interpret great masterpieces
satisfactorily by logical criticism alone. Apparently,
imagination follows certain personal paths before it
produces results in visible shape. Initially conceived
images have then to be developed and translated into
concrete form. Many things can happen to such 'inner
images' from their conception until their realization.

Artistic ideas follow certain trends and depend on
circumstances which can be summarized in the
following way:
  (i) the historical periods, the social conditions and
       aesthetic trends in which they were conceived;
 (ii) the creative imagination and talent of the artists
       active in such periods;
(iii) the tradition of any particular race or people (of
       great importance as this obviously involves aesthetic
       inheritance and tradition).

Sources of inspiration
The unconscious mind obviously plays a major role in
the development of new artistic ideas; the initial driving
force behind any major art inspiration could be regarded
as outside the conscious control. Painters such as
William Blake, Odilon Redon and Samuel Palmer
produced highly visionary paintings, the content of
which probably stemmed from unconscious imagery.
Redon wrote of the importance of the unconscious in
the creative process: 'Nothing is achieved in art by will
alone, everything is achieved by docile submission to
the advent of the unconscious.' Blake asserted that 'the
real centre of knowledge is the inner consciousness'.

This inner consciousness was the object of the new

9 *The Battle of the Sea Gods* by Andrea Mategna. The combination of mythology and the scientific study of anatomy as expressed by the early Renaissance artist
By courtesy of the Victoria and Albert Museum, London

science of psychology at the beginning of the century and it was not surprising that, at the same time, artists were involved in research of their own imaginative sources. Paul Klee wrote of seeking 'the truth behind exterior surfaces'. He instructed that the student 'follow the ways of natural creation, the becoming, the functioning of forms'; and so he would 'become like nature itself and start creating'.

During the 1920s the words 'stream of consciousness' and 'chance' described much of the activity of artists

10 *Comedy* by Paul Klee. The search for truth behind exterior surfaces
By courtesy of the Tate Gallery, London

such as the Dadaist Jean Arp, the Surrealists Ernst and
Dali, poets such as Breton, Eluard and Aragon. They
undertook to eliminate as far as possible any logical
control of their work and to act on unimpeded impulse.
The American painter, Rothko, said more recently that
he wished to 'remove all obstacles between the painter
and his ideas . . . because memory, history and
geography merely complicated the issues'.

Historical periods
Human perception depends on a complex of cultural
and psychological factors. Investigations into the evolution of visual perception show that the history of art

forms is closely linked with human behaviour patterns, which in turn give rise to perceptual behaviour. Within each cultural period similar kinds of changes have taken place over long periods of development. The process of 'seeing' in itself has presumably remained the same; what has changed is the interpretation and rendering of the seen images. Just as one can speak now of a historical evolution of perception because new ways of seeing are conditioned by the historical periods which produced them, so the way in which space is conceived is a historical phenomenon. Francastel puts it as follows: 'the conception of space is an expression of a specific type of relation between man and his environment'. Each society developed its own artistic conceptions, which influenced all persons living within that particular historical period.

In Pre-Renaissance art the prevalant conception of space was two dimensional. Artists were concerned with depicting devotional images, rather than the space around them. It was accepted as self-evident that shapes and forms in the upper picture plain were supposed to be further away from the spectator than those in the lower half. Even today many Eastern people continue this tradition of two-dimensional representation — their art remains one of symbols and patterned surfaces.

11 Pen and ink drawing by Rembrandt, expressing intense human emotion by
deceptively simple means
By courtesy of the Victoria and Albert Museum, London

With the Renaissance a completely new form of
rendering space was evolved. Leonardo da Vinci was
one of many to investigate and explain the idea of
perspective. Such investigations ran parrallel with, and
were probably stimulated by, an intense interest in man
for his own sake rather than as an attribute of the divine.
These principles were applied to create a pictorial
representation of three-dimensional space and are still
widely used today.

Leonardo's artistic curiosity was matched by scientific
research and his notebooks abound in details from the

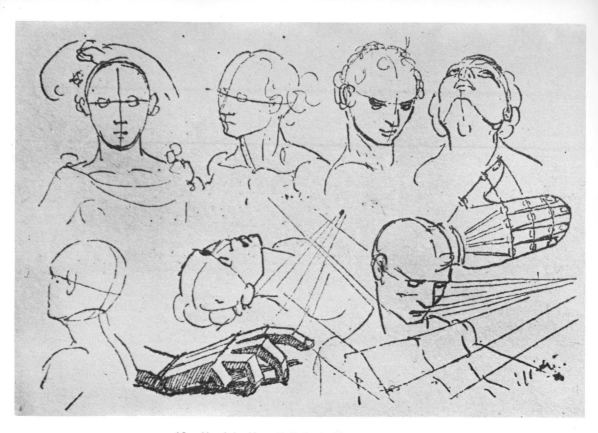

12 *Heads* by Hans Holbein the Younger
From *Handzeichungen von Hans Holbein dem Jüngeren* by Hans Ganz.
Berlin 1908

mathematical shapes of nature. This desire to understand
and use the shapes underlying all forms in nature can
be traced throughout the history of art. Most great
artists in the past were deeply interested and concerned
with abstract shapes. Some of the drawings by Hans
Holbein are proof of deliberate research into basic
shapes. Even his portrait paintings and studies take
account of such shapes, and in his drawings they
dominate everything else. Rembrandt, too, finally
arrived at a graphic expression which embodied all
underlying shapes with the utmost force. This almost

13   Movement studies by Bracelli

analytical approach did not in any way prevent him from
expressing the most intense human emotions.
Equally interesting are Bracelli's movement studies,
made in the seventeenth century.

Cézanne, two centuries later, became increasingly
absorbed by his research into the shapes underlying
all objects in nature. His study of plains and volumes
led to a new approach to problems of space and their
application. The cube, the sphere and the circle were
recognized as formal components of objects and were

14 *Mont Sainte-Victoire* by Paul Cézanne
By courtesy of the Phaidon Press (George Allen and Unwin Ltd)

used to build up balanced and integrated compositions.
'Nature', said Cézanne, 'is more depth than surface'.
The Cubists took Cézanne's investigations of space
much further. They wished to record this depth in
detail and in doing so projected different aspects and
angles of space on to one picture plain. By this shift of
observation the paintings acquired a pictorial freedom of
composition.

'Every cultural period has its own conception of space, and it takes time for people consciously to realize it', wrote Moholy-Nagy.* This consciousness has been realized with very different results in the twentieth century. Mondrian reduced shapes to a minimum of verticals and horizontals and based a whole philosophy on this conception. Kandinsky tried to penetrate and analyse the visual means of expression, following, as he put it: 'the necessity to break through to the interior'. He called the inner necessity the 'dominant rule of art'. Kandinsky's influence has been widely felt: partly owing to his teaching at the Bauhaus (1922–33), but also due to his following a deliberate line of research – expressed in his books *On the Spiritual in Art* and *From the Point and Line to the Plain*.

A number of artistic modes have been evolved in this century in order to reach a deeper sense of pictorial meaning. Activists such as Jackson Pollock, in the 1940s and 1950s, achieved amazing results by direct application of splashed colour patches – but even in this case not all conscious control could be completely elimated although such artists might not have been aware of this fact. A similar starting point is provided

* Moholy-Nagy *The New Vision (The Documents of Modern Art)*. New York: George Wittenborn Inc. 1947 (first published 1928).

15 Early 'concrete' woodcut (1913) by Kandinsky, illustrating his analytical
approach to his poems and graphic work of that period
From *XXe Siècle* no. 3, 1938

by mechanically produced doodles — using such scribbles as the initial starting point eliminates almost
entirely direct control over the shapes.

Present tensions, fears and hopes have produced a
totally new kind of pictorial language and a more
conceptual view of art, much of which bears direct

16　*Goat's Skull, Bottle and Candle* by Pablo Picasso
By courtesy of the Tate Gallery, London

comparison with recent scientific discoveries. This can
be seen in particular in the kinetic art of the 1960s,
which probably represents the most complete interaction of art and science to date; present experiments in
art and technology continue to stimulate further
questioning of the established boundaries of creative
activity.

3. Art forms in nature

The art critic, Clive Bell, has written: 'The quality of all great masterpieces of art, irrespective of their origin, lies in their significant form. In such forms, lines and colours combine to create shapes which stir our aesthetic emotions. Such qualities are common to all great works of art. Here, every element is correct and necessary and cannot be disposed from the place where it stands.' Any scientific investigation into organic or inorganic structural development shows that many purely functional problems, arising during long periods of evolution, have led to shapes and structures which provide a perfect solution from an aesthetic point of view. This applies not only to visible shapes, but also to the fundamental organization of formative particles of matter such as atomic and molecular structures, as well as to all organic cell structures. It is wrong, however, to assume that all formal problems solved by nature can be equated with similar problems arising during aesthetic or artistic processes.

D'Arcy Thompson, in his remarkable book *Growth and Form* (first published 1917), investigated and illuminated many outstanding biological problems. He was particularly successful in his research of certain elementary forms such as the growth of spiral shapes in shells and horns. Recent research into genetics and

17 The fruit of a palm tree growing in the Seyschelles, called 'Coco de mer'. The same fruit serves as a symbolic emblem of the *yari* (vulva) of the goddess Parvati (companion of the god Shiva) in the Indian Tantra culture. Tantric Buddhism incorporates Hindu and pagan elements

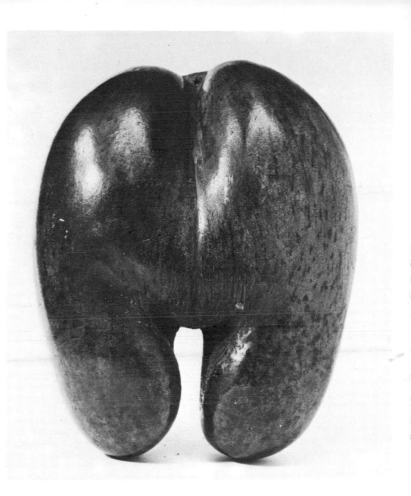

hereditary science was, of course, unknown to Thompson — nevertheless, his work revealed many aspects of related forms in nature. He pointed out that closer analysis of natural phenomena through physical or mathematical research made their beauty even more apparent. Such thoughts seemed alien at a time when beauty in nature was considered to be based almost wholly on its undisclosed and unsolved mysteries.

49

18 Architectural domes
From *Architects and Architecture, New Directions in America*. Paul Heyer, Penguin Press, 1967

Recent understanding of living building materials such as proteins and nucleic acids has revealed important facts concerning the hereditary, genetic codes and, consequently, the development of cellular structures. Such findings may be compared with the solutions to certain problems arising in the field of creative art. In many cases the amazing similarity of solutions is very striking. Thus Pollock, in some of his pictures, arrived at compositions of pigment particles in motion, very similar to the images caused by sound-wave agitation of liquids such as glycerine.

Other, and possibly more concrete, examples are to be found in certain viruses, which arrange their units in such a manner that their final structures appear as many-cornered domes when observed under an electron-microscope. Recent research has revealed this to be the only way in which these units can find and

19   Tobacco leaf virus
From *Module, Symmetry,
Proportion,* edited by G. Kepes.
Studio Vista and George
Braziller 1966

reach a solution to ensure their multiplication, which In turn leads to an exact reproduction of their nucleic acid units.

Similar shapes have provided the solution, from many points of view, to problems in structural engineering and architecture. By arranging a great amount of mass units, dome-shaped structures achieving a stable and harmonious balance have been constructed.

The building block of all higher organisims is the cell. Predetermined in their disposition by the heritable code, the cells multiply to form cell complexes and cell structures. Most cellular structures arise from cell division, known as 'mitosis' which forms one of the most complex of all movement patterns in nature. This process, vital to the maintenance of life, involves the transmission of the heritable nucleic material to the new

cells and follows definite phases in which the chromo-
some threads move from the centre of the cell to
opposite poles within the cell, while a dividing wall
forms itself, thus creating two new daughter cells. Each
of the new cells possesses the same heritable units
(genes) as the mother cell. The phases follow almost
rythmic sequences comparable to some ancient
cultural dance.

Some basic patterns of natural growth have features of
outstanding beauty, harmony and balance. Such shapes,
irrespective of their purpose, may be grouped together
from a purely formal and aesthetic point of view, in no
way implying any scientific classification. Where strictly
geometrical forms are involved, however, it is possible
to be more specific for certain forms follow lines
exactly coinciding with the results of scientific research
into similar problems.

Shapes and varied patterns, though appearing at
different times and developed for completely different
purposes, often show great similarity. Such fundamental
shapes reoccur both in organic and inorganic formations
and also in the transitionary processes of development.

At the same time the arrangement into different patterns

20   Senuffo painting. The Senuffo inhabit the Upper Volta (formerly the French
     Sudan). The beautiful designs are produced by a hot knife, dipped into dyes
     and applied to hand-woven material
     Author's collection

can be observed even at the most basic levels of all
matter. The calculable electron wave patterns in atoms,
the most basic form of matter, show an astonishing
form of varieties of shapes. Such wave patterns determine, in fact, the larger structures of atomic arrangements.

Interest in the shapes and forms existing in nature was
almost non-existent in Western society before the
Renaissance. So-called 'primitive art' showed much

more awareness of existing natural art forms than the art of more civilized societies of the past, which were less involved with the environment. In this respect it is interesting that it was only through the liberation of abstract art that the implicit harmonic pattern of formal art was made explicit to us. We now live in an age uniquely conditioned to finding direct aesthetic satisfaction in the geometric and organic patterns of nature.

Fundamental patterns
It is said that all matter evolves in definite patterns which are dictated by the fundamental arrangement of the particles which form them.
Erwin Schroedinger, in 1926, examined the behaviour of electron waves when confined to a nucleus. He calculated their frequencies and their most characteristic patterns. The result of the relationship of the electron wavelength and the electron velocity appears in a series of vibrations. A connection had been found between the existence of atom states and the wave nature of the atom. Schroedinger succeeded in establishing the electron wave picture of the atom and found that the vibrations of electron waves correspond to the energies of absorbed 'quantum states' of energy (Plank). It is obvious that inner atomic wave patterns cannot be

directly observed, but their frequencies and extensions may be measured. From calculations, therefore, it is possible to construct models showing electron wave patterns in successive quantum states of electrons, confined by a nucleus in an order of increasing energy or frequency. With higher frequencies, the pattern becomes increasingly involved. The lowest state, namely the electron wave picture of the hydrogen atom, where the nucleus confines one single electron, is the simplest and shows spherical symmetry. Higher frequencies show much more complex forms. The ninety-two elements, making up all substances in existence, consist of basic matter which gives each its particular identity. The electron wave pattern thus determines the build up of different materials and their patterns change in accordance with increasing energy. The fundamental forms on which all matter is built depend on these patterns.

The pattern of life
It is now known which kind of evolving, complex molecules will finally produce living forms, crystalline shapes or any other state under the influence of ever increasing temperatures. In living cells, two kinds of large molecules are found, namely the protein and the nucleic acid molecule. Each living cell inherits all

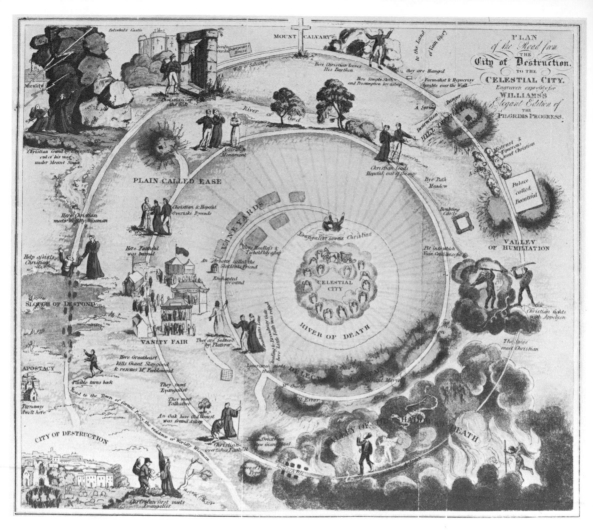

21 'The City of Destruction' showing spiral formation
From *The Pilgrim's Progress*

possibilities of the parent cell. During cell division, all
cell elements are doubled and it may pass all inherited
factors on to the two emerging daughter cells. The
most important feature of cell division is the doubling
of all molecules which contain the 'genetic code' of

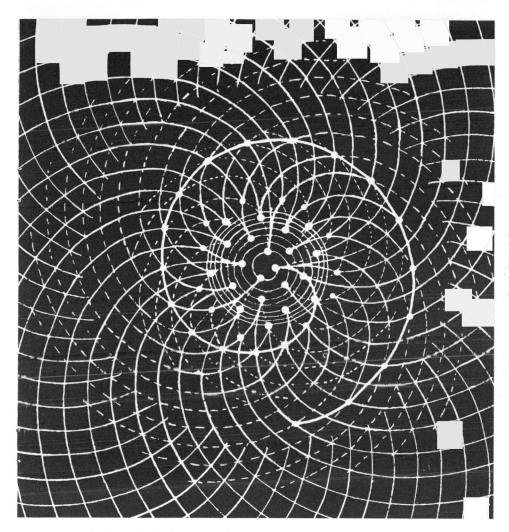

22 Example of plant growth in spiral stages, depicting the complicated spirals
evolving during growth in a composite flower such as a single daisy. Each
separate floret in the capitulum forms a crossing point in the evolving spirals
From *Croissance et Topologie* by André Hermant. Pragma 1971

each cell. The bearers of genetic information seem to be
the nucleic acid molecules. These are the Deoxyribonu-
cleic Acid (DNA) and the Ribonucleic Acid (RNA)
molecules. The complex DNA molecules are made up
of many hundreds of thousands of simple sub-units

23 A geometrical staircase, The
Queen's House at Greenwich,
by Inigo Jones *c.* 1635
From *Illustrated Glossary of
Architecture* (1966) Faber
and Faber

24 Sea shell

called 'nucleotides', which contain the nitrogeneous sub-units or bases. They appear in pairs in an enormous variety of combinations. The possible characteristics of all living organisms passed on from generation to generation seem to be contained in them. The two DNA nucleotides are twisted into a spiral ladder. This double helix can truly be called 'the spiral of life' for on it depends the maintenance of life in all living cells.

The DNA helix bears coded rungs or bases. Only two code units form one rung in the 'ladder' and their exact order conveys exact genetic instructions. There are four such bases (or code units), namely: Adenine, Guanine, Cytosine and Thymine (called A, G, C and T for short). These bases invariably pair so that A always links to T and C to G. The four bases are joined to each side of the 'ladder', made up of phosphate and sugar groups. In a dividing cell, the DNA is able to duplicate itself by a splitting of the DNA helix and then produces an exact copy of itself. This happens because it attracts identical chemical units to the exact points where they are needed, to replace the missing half of the ladder. There are now two new spirals, each identical to the original one.

25 Model of the molecular
    structure of the DNA spiral,
    forming a double helix
    By courtesy of the British
    Museum (Natural History),
    London

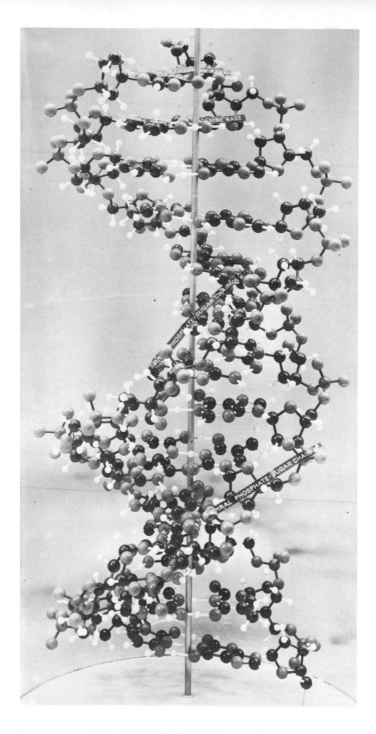

The DNA produces the molecular organization of the proteins by synthesis. Proteins are made up of amino-acid molecules which are arranged in chains and of which there are twenty kinds. All possible kinds of proteins can be synthesized through innumerable combinations. The sequence of nucleotides in the nucleic acid, containing the heriditary message, is thought to determine the sequence of aminoacids in the protein, which it manufactures within the cell. A transmission of the genetic information from the DNA (in the chromosomes) to the places in the cell, where proteins are synthesized, occurs by means of the RNA.

The energy required for these and similar vital processes is always given by Adenine Triphosphate (ATP) which is produced by the protein molecule from sugar. It is not possible to enlarge here on the fantastic chemical organization which exists to maintain and renew all living cellular shapes.

Recent findings concerning the heritable code (genes) and the knowledge acquired about how living materials are built up chemically must be regarded as being one of the great discoveries of this age.

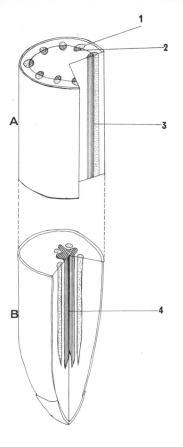

26 Transverse section and
    longitudinal section
    through a stem (A) and a
    root (B) of a higher plant.
    The skeletal elements:
    1. Bast
    2. Wood in ring form
    3. Longitudinal section shows
       the position of the bundles,
       running the length of the
       stem
    4. In the roots the skeletal
       elements are clustered in
       the centre for strength

    To the right:

27 Leaf skeleton showing mid-rib
    and veins

Structural plant forms and function
Certain structural problems evolved by higher plants
demonstrate very clearly that no structural engineer or
artist could have found a more adequate solution,
either from a practical or from an aesthetic point of
view. Stress problems were solved by plants and have
served as prototypes for problems arising in the
structural design of modern buildings. The internal
structure in trees and other higher plants shows the
existence of the 'central cylinder', which consists of
vascular bundles containing conducting tissues. These
are arranged in a circular shape within the outer cell

structures. The double function of such cellular structures is well known. Firstly, they allow for the transport of liquids such as water and mineral salts to rise from the roots to all parts of the plant which are wood. Secondly, the dissolved sugar made in the leaves descends in the phloem (bast), forming part of the vascular bundles. The third function, however, is that, because these compact bundles descend the whole length of the plant, they can be regarded as its skeleton. The arrangement in ring form, which will remain thus throughout their lives in spite of yearly thickening, is the best possible solution to stress problems arising from wind action. The study of transverse sections shows not only that arrangements in ring form are highly satisfactory from the practical point of view, but that they are also the most aesthetically pleasing (fig. 26).

A further point of interest is that these (skeletal) stress elements change position completely once they enter the roots. Here the ring shape disappears and they are centred in the middle of the root, which again is the best position they could possibly occupy to resist uprooting by wind pressure. Modern structures in steel and concrete are based on almost identical arrangements of their components.

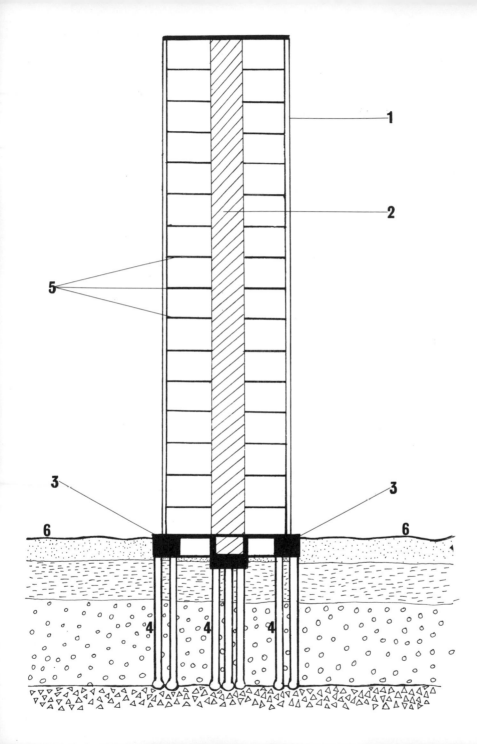

28 Cross-section of a typical
    high block founded on piles
    1. Reinforced concrete façade
       taking vertical loads of
       floors
    2. 'Core structure' takes
       vertical loads from floors
       and all horizontal wind
       loads
    3. Reinforced concrete pile
       caps
    4. Reinforced concrete piles
       carrying building loads
    5. Floors
    6. Ground level

Figure 28 shows a cross-section of a typical high block
founded on piles. Many features in this kind of building
are strongly reminiscent of the organization of tree
structures and their function. The so called 'core
structure' in the centre of the building consists of
reinforced concrete and is giving the firm stability
required and, at the same time, taking the vertical loads
from the floors as well as all horizontal wind loads. This
core contains such service features as the stairs and lifts,
and here again a parallel can be drawn with functional
elements within the tree trunk. The underground structures, carrying the load of the building, consist of piles
deeply sunk into the soil. This is comparable to the
living root systems.

The plan of developing ferro-concrete as an element
of construction had already been applied by the famous
Swiss engineer, R. Maillart, as far back as 1900. He
was presumably the first person to construct a certain
type of concrete bridge and also to build unsupported
ceilings (mushroom ceilings) held up only by pillars
but otherwise self-supporting. The reinforced concrete
contained iron rods which were set into the concrete,
running in two directions.

A very wide range of different examples demonstrate similar structural solutions at different levels of plant evolution. The vital organ for the manufacture of food by photosynthesis is the green leaf, which requires the maximum amount of sunshine, without which sugar and starch production would be impossible. The same skeletal elements as are in the stem and branches continue into the leaves as veins, which keep them firmly spread and exposed to the sun. Again the double functions of supplying food and water and acting as skeletal elements are combined. The leaves are furthermore capable of movement and adjustment, receiving a maximum of sunshine. Consequently they are arranged in patterns in such a way as to fulfil the laws of balance and harmony. As will be seen later, all organs in plants change according to their environment and function.

Growth and rhythm
The growth and development of young plants from seeds is impressive to watch, particularly if photographed in speeded-up sequences. Apart from the purely biological facts, one is immediately struck by the exact rhythm of growth. A very definite sequence of growth development can be observed, even with the naked eye in tropical plants. The folding and unfolding

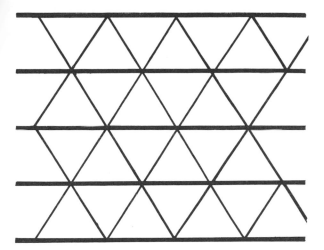

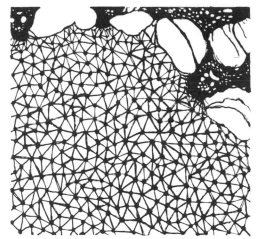

29 Tubular roof structure

To the right:
30 The structure in the stem of a rush. This aquatic plant breathes under water. The air is stored in the intercellular spaces

of leaves and petals follows rhythmical patterns dictated by daylight. Movements in climbing plants show extraordinary twisting convolutions caused by the support around which they wind.

Ecological factors will produce completely different shapes and forms: for example, the survival of plants in alpine or sub-Arctic conditions forced the appearance of hairy, protective covers similar to those of some animals. Tropical plants, on the other hand, developed leaves and flowers suitable to the climate and plants growing in arid zones acquired protections against the loss of water. Aquatic plants acquired air chambers in their tissues, allowing them to breathe under the water and to float on it. All these different adaptations and cellular organizations form structural patterns of great harmony and beauty.

Symmetry
The beauty of some lower animals and plants is mainly due to the symmetry of their shapes and structures. This is particularly striking in some invertebrate species such

as the Medusae, or starfish, where all organs are arranged according to a planned 'radial symmetry' — all parts of the body are arranged around a central point so that any sectional planes passing through the centre divide the animal into two exactly equal halves. In animals built on the principle of 'bi-lateral symmetry', on the other hand, the parts of the body are arranged on the right and on the left side which only one plane would divide into two equal halves. No other plane will divide the body into two halves which are alike.

Even at low levels of organic evolution, extraordinary symmetrical shapes have developed. The minute single-celled creatures belonging to the 'Protozoa' such as the Foraminifera or the Radiolaria, or else the Diatoms which are plants, construct shells from chalk and silica of exquisite beauty. They show perfect symmetry and are often of most intricate structure. The many varieties of minute skeletons, often in hexagonal shapes, were investigated by many scientists, including D'Arcy Thompson, who thoroughly described the hexagonal structures of Radiolaria. It has now been found that even more intricate skeletal structures underlie, in some cases, the outer basketwork of some one-celled organisms. It seems extraordinary that such minute creatures should develop such complicated frameworks.

The symmetry of leaf and floral distribution on the stem is well known. Floral symmetries depend very much on the type of plant: the sunflower, for example, carries all its florets, which evolve in a spiral shape, in one cluster concentrated in the umbel (fig. 22). The prevalence of the spiral shape in art and architecture reflects man's preoccupation with beauty in Nature. The spiral arrangement of green leaves along the stem was studied by Leonardo da Vinci, who was very much aware of repetitive shapes, as his study of water movement reveals.

Functional patterns
The forms of organisms are a direct consequence of growth and it has now been stated that it might be more correct to say: 'for what purpose did it happen?' rather than: 'why did a particular process happen?'. The best solutions to environmental problems have assured the survival of living species, which had to evolve certain organs in order to maintain their existence.

Both in plants and animals repetitive, though not always strictly symmetrical, arrangements of features are often

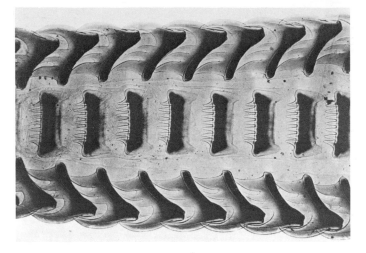

31 Magnified photograph of a snail's palate showing the repetitive structure of the teeth which act as an endless moving band

found in structures which serve specialized functions. The positioning of the taste buds in higher animals clearly shows a regularity, and so does the palate of the snail. The teeth sitting on the centre part impale the food which is shredded by the side teeth against which it is pressed.

Another example of functional arrangement can be found in the structure of the insect eye, the same type also being found in some crustaceans. Many separate optical units are arranged in each eye to form the regular, but intricate, structure of the so-called 'compound' eye. Here a number of elements called 'Ommatidia' form each eye and each of these lenses is capable of forming a separate image. The separate images form in combination a 'mosaic' image of the whole visual field. The eye is controlled by a pigment within it and, in strong light, this pigment spreads through the cells, isolating each lens, each forming a separate, small but sharp image of part of the viewed

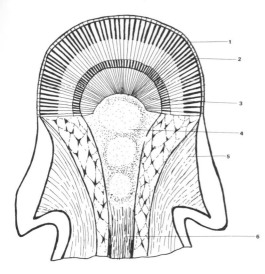

32 Diagram of an insect eye. The longitudinal section shows the internal structure
1. Corneal lenses
2. Ommatidia (lenses)
3. Nerve fibres
4. Optic ganglion
5. Eye muscle
6. Optic nerve

To the right:
33 Wing scales of a moth. Vertical ribbing, base attached to wing skeleton

field. In weak light, however, the pigment retracts and the whole eye together renders a single but diffused image (fig. 32).

Pollen grains of higher plants often show symmetrical patterns 'engraved' into the very tough outer cover. These characteristic designs are typical for each plant family which produces them and the outer cover, being indestructible and non-rotting, can preserve the pollen almost indefinitely (examples have been found in peat bogs where they had lain preserved for centuries).

The scales developed by butterflies and moths are attached to the wing skeleton, forming varied patterns according to the species. The scales show vertical ribbing and each has a base which attaches it to the wing skeleton, where they are beautifully arranged with geometric precision (fig. 33).

Equally impressive is the arrangement of the protective scales covering the bodies of fish, adapted to their respective cylindrical shapes. The infinite variety of striped and spotted designs found in sea shells can be regarded as a means of protection, since it is much more difficult to discern and discover such 'camou-flaged' shells. The vast field of protective patterns and forms, which is necessitated by dissimulation (mimicry) is of great interest. Another example of repetitive structures can be seen in the nests constructed by social insects such as bees and wasps which both live in colonies. Inside the hive are numerous hexagonal cells made from wax produced by the worker bees. Since antiquity these cells have been studied and admired.

The many existing species of corals are sea polyps. They live mostly in colonies, forming for their protec-tion strong skeletons of calcium carbonate extracted from sea water. They multiply mainly by budding, and thousands of such minute animals formed the famous coral reefs of the South Pacific. The whole colony is internally connected, in some ways comparable to the canal system of the sponges. Regular domes are formed by certain corals through symmetrically budding in all directions.

34 Coral colony

35 Coral skeleton

Crystalline shapes

Crystals grow by building up surface layers in a regular symmetrical manner. Their shape is determined by the geometric arrangement of the atoms which compose them and they are of a definite chemical composition. Each chemical combination forms a typical type of crystal shape, always producing exact proportions of crystalline faces and exact regularity of angles. It has been possible, therefore, to organize the types of existing crystals into six definite growth groups, each of which repeats all the time the same formations. Crystalline shapes are quite stable in spite of their comparatively large size and they appear at the lowest level of the quantum ladder, previously discussed, where matter is established at very low temperatures. When cooled sufficiently, nearly all substances can crystallize and the formative particles are arranged in a total order, which then assumes complete immobility. Similar large macro-molecules are combinations of many ordinary molecules which were assembled under special conditions. Similar molecules are also responsible for the appearance of life (see DNA spiral). The internal pattern of atoms is reflected in the outer crystalline shapes. It is therefore possible to discern the expression of internal structures from the outer visible shapes, which in each case is regular and constantly repeated.

Each chemical compound produces its own typical crystal. Such crystalline forms remain always the same, provided the temperature remains constant.

Crystalline structures can now be determined with the help of x-rays. It is possible to decipher the atomic structures of minerals by the reflections from the electrons that form the outer shells of the regular arrangements of atoms in the crystal.

Cleavage is a very distinctive property of most minerals, which will always break along precise planes parallel to each other. Breakage can take place in two or three directions of cleavage, which in each case is geometrically repetitive, depending on the relative arrangement of atoms. The world of crystals and the intrinsic beauty of clear-cut crystal formations is extremely attractive in all its expressions. The repetition of definite, angled shapes often enhanced by beautiful colouring is most appealing and gratifying.

Fossils
Fossils found in stratified rocks indicate the geological time periods in which they existed. Thus they became a very important clue to evolutionary change. The most astonishing fact is that some petrified animals seem to

37 Ammonites. These fossil
shellfish became extinct 100
million years ago. The fossil
on the right is actually the
rock impression of the
petrified specimen on the left

38 Sea urchins: one is petrified,
the other is the shell of a
present day species. The long
spines covering the living
animal have disappeared

39   Petrified fish skeleton

have developed lasting shapes which still appear today. This is the case, for instance, with petrified Foraminifera, which formed the chalk cliffs and other sedimentary rocks; or Diotomite, a shale which includes miriads of skeletal remains of the Diatoms. These single-cell plants still live in uncounted millions in the surface waters of the seas. Their average size is about 50 microns (or 2/1000ths of an inch). They appear in eliptical, triangular, or rounded shapes with intricate internal skeletal structures.

The fossil Ammonite (a molusc) lived in spiral shells for 125 million years of the Mesozoic age. This marine animal became extinct some 100 million years ago and the Ammonites were very similar to the present day Nautilus of the tropical seas. The shell of the Nautilus is of great interest because it is an absolutely accurate equi-angular spiral in which each radius whorl represents a growth from a previous radius. In growth there is a constant percentage increase in size and the

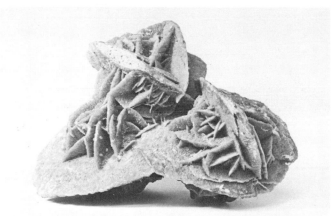

40   Crystals exposed to the action of desert wind erosion.
Barytes crystals, forming the so-called 'Desert Rose'.

resulting complete shell will be in the form of a logarith-
mic spiral. Each whorl is about three times the breadth
of its predecessor.

Erosion
Weathering and erosion are responsible for many of the
changes affecting the earth's crust. Stream erosion,
desert and sea erosion all helped to change the surface
of the earth. Rocks are pounded and ground down by
the waves year after year on rocky sea coasts so that
the resulting cliffs and boulders take on extraordinarily
regular shapes. The spherical sculptures of such artists
as Brancusi, Moore and Hepworth, reflect the influence
of such forms.

Desert erosion produced enormous arid areas in Africa,
Asia and elsewhere, submerging and changing the
whole landscape. Similar to the intricate wind patterns
found on the surface of sand dunes, the receding water
leaves identical ripples on the beaches at low tide. The

79

grinding action of wind and water produced ever
smaller and smaller units, which at some stages were
again pressed into firm, sedimentary rocks. Crystalline
shapes of great beauty were sometimes produced by
the blasting of sand on to stones and rocks with hard
inner cores. The inner core resisted the action of the
wind and appears now in strange crystalline shapes.

Movement patterns
The influence of the magnet on metal filings is well
known. These filings will follow the streamlines set up
in a magnetic field by the positive and negative poles
of the magnet. Moving the paper which covers the
magnet and on which the metal filings are placed, will
also alter and move these filings until they re-form
themselves. In the same way, electricity produces two
poles where the negative electrons in matter are
attracted by the positive centres of the atom. An electric current in a spiral wire acts, therefore, like a magnet
creating a magnetic force. The study of magnetic and
electric fields led finally to the conclusion that the field
will propagate in all directions at the speed of light.
This, in turn, led to the discovery of the nature of light
and established the nature of all electro-magnetic
waves (see also Chapter 2).

Some simple examples of confined waves set up repetitive patterns of one kind or another. If a string is attached between two near points under tension, vibrations set up in it will produce a standing wave. The shape and frequencies of such waves are now determined and only such vibrations can be set up in which one, two, three, etc., half wavelengths fit in the space between the attachments. Most musical instruments are built on the principle of assigned, characteristic frequencies. String instruments or wind instruments follow these principles. In wind instruments the frequency of air waves enclosed in metal pipes produces the tones. Tonal patterns are similar to such wave patterns.

The close relationship between form and function is evident in the complexity of even the simplest organic structures. The fine, hair-like threads found in many one-celled organisms are called 'cilia'. Some simple protozoa use the cilia as a means of movement. These structures have been extensively studied under the electron-microscope and compared with other cells. Cilia are very small, hair-like structures, consisting of a stalk and a basal body which extends into the cell. The motion of each cilium involves an effective stroke, followed by a recovery stroke. They beat in a definite rhythm (a metronomal movement), which appears in successive waves, forming a repeating pattern. The source of energy required derives from the basal body and is the same ATP previously described. Their movement stops immediately they are separated from the basal bodies, which are arranged in a specific pattern of nine units.

The photo-sensitive cells in the retina of the human eye (the rods and cones) evolved originally from modified cilia. The structure of the basal bodies and the rootlet fibres in these cells confirm this. The light-sensitive parts of the cells are connected with the inner part of the cell by a stalk which transmits light impulses to the optic nerve and the brain.

4. Art and the unconscious mind

The history of science shows that the same discoveries are sometimes made simultaneously by different persons. Charles Darwin, for instance, developed his theory of the origin of the species at the same time as A. R. Wallace; they then joined forces to write a book on the subject. A number of such apparant 'chance' or 'coincidental' events are known. A possible explanation can be found in Jung's theory of 'synchronicity', or theory of 'meaningful coincidence'. In each case a new hypothesis of great importance has evolved simultaneously and completely separately. According to Jung, this happens when significant, unconscious archetypal forces are activated. Such a coincidence is not identical with a coincidence depending on causal connection. An accausal connection principle assumes that an unconscious knowledge links a physical with a psychic condition. If this is so, then events which appear 'accidental' are in fact physically significant.

The physicist, W. Pauli, discussed similar problems with Jung. Pauli was aware, for example, of the many parallels between nuclear physics and the laws concerning the 'collective unconscious'. He pointed out that our present ideas of evolution might be reassessed in the light of the relationship of biological processes with the unconscious mind. He believed that the investigation of outer objects should run parallel with psychological investigations of the 'inner origin' of scientific concepts. The revolutionary changes initiated by psychology superseded the classical conception of cause and effect. Modern nuclear physics discovered new worlds and laws similar to the laws governing the human mind.

In order to assess the significance of psychology a number of intimately associated ideas must be considered. The concept of symbols is as basic to our understanding of psychology as it is to our understanding of art. Symbols have a deeper and more complex meaning than signs, for they enable us to discuss abstract problems. *The Oxford Dictionary* defines a symbol as: 'A thing regarded by general consent as naturally typifying or representing or recalling something by possession of analogous qualities, or by association in fact or thought.' Symbols have always played an important role in religion and the visual arts, and throughout human history a close interplay of religion and art can be observed.

One of the most significant symbols used universally has been the circle, for the circle or sphere (mandala) expresses the totality of human nature and environment. In primitive, as well as more sophisticated, religions it is used as a symbol for the final wholeness. In Zen Buddhism it represents human perfection in the form of the lotus flower — a wholeness of Buddha and also of the god Brahma. The petals of the flower are divided by rays, signifying spatial orientation, symbolic of the need of psychic orientation of human beings. This is identical with the four functions of human consciousness enumerated by Jung, namely, thought, feeling, intuition and sensation, which enable men to comprehend and deal with impressions and experiences. Jung describes a symbol as: 'the best possible designation or formula for a relatively unknown element, which is nevertheless recognized as being present or recquired'.

In art, human feelings and emotions are expressed in symbolic form — colours and shapes are used to communicate meanings which can be conveyed only with difficulty in normal language.

Written words are representations of spoken sounds.
Alphabets developed originally from images as symbols
of language. Such pictorial alphabets were used by the
Egyptians and the ancient Sumarians. Their cuniform
scripts engraved into cylinder seals, were later rolled
and impressed into clay tablets. This type of picto-
graphic script is very expressive and originally formed
the basis of most alphabets.

41   Cylinder seal and impression *c*. 500 BC. King Darius of Persia in his chariot
      hunting lions. The inscription gives his name in three languages
      By courtesy of the British Museum

42 *Metamorphosis of Narcissus* oil painting by Salvador Dali
By courtesy of the Tate Gallery, London

The conceptions of symbolic expressions used in
modern analytical psychology need clarification. As a
rule known things are not in need of symbolic expressions, which are reserved for manifestations of the
unconscious mind. The psychoanalyst, Freud, regarded
the unconscious as being the personal unconscious
mind of each person. Jung, on the other hand, saw it
as consisting of the 'collective unconscious', underlying
the personal unconscious. The collective unconscious
would therefore be identical for everyone. One of the
principal aims of the movement called Surrealism was

87

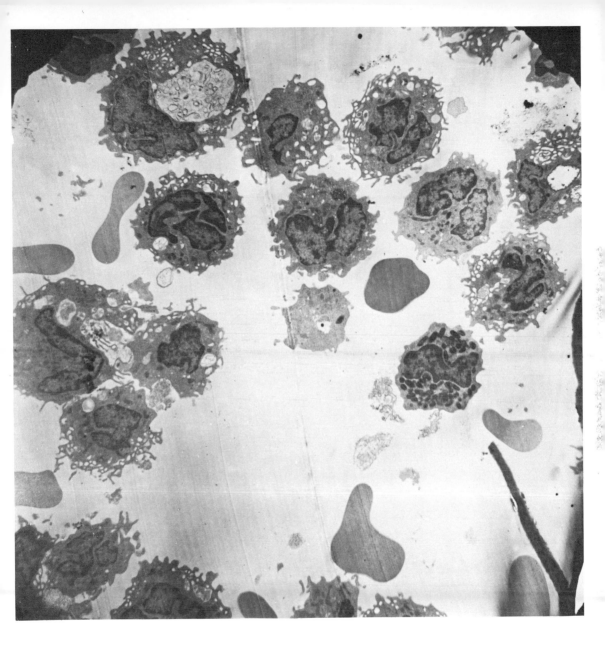

Opposite:

43 *Woman, Bird by Moonlight* by Joan Miro
By courtesy of the Tate Gallery, London

Above:

44 Cells of the human body, magnification × 4000
Courtesy of the University of Bristol (Department of Pharmacology)

to involve that part of the human mind which is not attainable by any conscious control. This new approach to artistic problems broke with all accepted traditions in literature and art by attempting to incorporate some of the findings of modern psychology in art expression. Much earlier, certain artists had tried to evolve images similar to those used in Surrealism, for instance, Hieronimus Bosch, or much later, Odilon Redon. But the Surrealists established a very close link with the latest findings of psychology. The movement, started by André Breton and Philippe Soupault, experimented with a method of 'automatic writing' used by psychologists. Breton tried to apply these methods to certain aspects of poetic expression. The imagery appearing in dreams and its symbolic meaning was of deep interest to Breton and his friends. Similar ideas were soon applied in the visual and pictorial field by an ever-increasing number of artists who realized the great possibilities of artistic expression in Surrealism.

Soon the control of reason in artistic productions was more and more disclaimed and everything was left to imagination and dreams which were consciously used as the gateways to an unconscious world. The influence of these ideas on the succeeding generation of artists was enormous. They all tried to live up to the one great surrealist maxim, to express 'the marvellous which is beautiful'. Each tried to give life to symbolic images as he perceived them.

The dream world became a world of reality. Film makers as different as Buñuel, Cocteau and Fellini, took this up with avidity.

The pictures painted by people suffering acute mental disturbance often reveal significant symbolic meanings, emerging from the unconscious mind. In such people, their conscious thoughts are almost obliterated by a stream from the unconscious regions of the mind. Such pictures often reveal distortions and a fragmentation of shapes and forms; the influence of such pictures on some modern artists is obvious. There is, however, a very profound difference between such images produced deliberately by artists and those produced by psychotics.

Art and science today have combined to produce a therapy for the mentally sick. So-called spontaneous painting and modelling provide a creative outlet for psychotics. The benefit and the liberation of this art therapy is due to the fact that it is not controlled.

Art in science reveals itself in many and more obvious ways, as for example, under the microscope in the structural patterns of living tissue. Science in its turn has provided a backbone to much of Western art. This can be shown in Leonardo's preoccupation not only with mechanical contrivances but also with the mechanics of various drawing systems. It can be seen in the pioneer work of Duchamp, Gabo and Moholy-Nagy in kinetic art, the language of movement. It is a paramount feature in Vasarely's preoccupation with optical illusion and so on; yet conversely, art may still play a most important role and be the key to the solving of many scientific problems. When Einstein said: 'Beauty is the first test', he was only reiterating the craftsman's remark that 'What looks right is right.'

Bibliography

Anderson, D. A. *Elements of Design*. London: Studio
Vista.

Argyle, M. *The Psychology of Interpersonal Behaviour*.
Harmondsworth: Penguin Press 1967.

Arnheim, R. *Art and Visual Perception*. London: Faber
and Faber 1967. *Towards a Psychology of Art:
Collected Essays*. London: Faber and Faber 1967.

Bandi, H. G. and J. Maringer *L'Art Préhistorique*. Paris:
Editions Holbein; Basle: Charles Massin & Cie.

Barbu, Z. *Problems of Historical Psychology*. London:
Routledge and Kegan Paul 1960.

Bertram, A. *William Blake*. London: Studio Publications
1948

Borradaile, L. A. (M. B. Yapp) *Manual of Elementary
Biology*. Oxford University Press 1961.

Burckhardt, J. *The Civilization of the Renaissance in
Italy*. London: Phaidon Press 1965.

Bury, J. B. (Ed.) *The Cambridge Ancient History*.
London: Cambridge University Press 1927.

Chardin, P. T. de *The Phenomenon of Man*. London:
Collins 1961.

Combe, J. *Jerome Bosch*. London: Batsford 1946.

Darwin, Charles *Origin of the Species* 1869.

Daucher, H. and R. Seitz *Didaktik der bildenden Kunst*.
Munich: Don Bosco Verlag 1969.

Dirac, P. A. M. *Principles of Quantum Mechanics*.
Oxford University Press 1958.

Ehrenzweig, A. *The Psychoanalysis of Artistic Vision and Hearing*. London: Routledge and Kegan Paul

Eysenck, H. J. *Fact and Fiction in Psychology*.
Harmondsworth: Penguin Press 1968 (first published 1925).

Fordham, F. *An Introduction to Jung's Psychology*.
Harmondsworth: Penguin Press 1953.

Freud, Sigmund *Totem and Taboo*. London: Routledge and Kegan Paul 1950 (first published 1913).

Gans, P. *Handzeichnungen von Hans Holbein*. Berlin: J. Bond 1906.

Gaunt, W. *The Aesthetic Adventure*. London: Jonathan Cape 1945.

Georges Michel, M. *Les Grandes Epoques de la Peinture Moderne*. New York: Brentanos 1945.

Ghiselin, B. *The Creative Process*.

Gombrich, E. H. *Art and Illusion*. London: Phaidon Press 1962.

Gore, F. *Abstract Art*. London: Methuen & Co. 1961.

Gregory, R. C. *Eye and Brain*. London: Weidenfeld and Nicholson 1966. *The Intelligent Eye*. London: Weidenfeld and Nicholson 1970.

Grove, A. J. and G. E. Newell *Animal Biology*.
University Tutorial Press 1966.

Hauser, A. *Philosophy of Art History*. London: Routledge and Kegan Paul 1958.

Hermant, A. *Croissance et Topologie*. Paris: Pragma 1971

Huyghe, R. *Larousse Encyclopedia of Byzantine and Medieval Art*. Lonfon: Paul Hamlyn 1969.

Jung, C. G. *Man and his Symbols*. London: Aldus Books 1964.

Klee, Paul *On Modern Art*. London: Faber and Faber 1948.

Lyddiatt, E. M. *Spontaneous Painting and Modelling*. London: Constable 1970.

McKellar, P. *Experience and Behaviour*. Harmondsworth: Penguin Press 1968.

Medawar, P. B. *The Art of the Soluble*. London: Methuen 1967. *Induction and Intuition in Scientific Thought*. London: Methuen 1969.

Naumburg, M. *Schizophrenic Art, its Meaning in Psychotherapy*. London: Heinemann 1956.

Peacock, C. *Samuel Palmer: Shoreham and After*. Cambridge University Press 1970.

Plokker, J. H. *Artistic Self-Expression in Mental Disease*. London: C. Skilton 1964.

Putnam, W. C. *Geology*. London and New York: Oxford University Press 1965.

Read, Herbert *Art and Society*. London: Faber and Faber
1945. *Education Through Art*. London: Faber and
Faber 1958.

Richter, I. A. (Ed.) *Selections from the Notebooks of
Leonardo da Vinci*. Oxford University Press 1952.

Schmidt, G. and R. Schink *Form in Art and Nature*.
Basle: Basilius Press 1960.

Seagar, A. T. 'The Surface Structure of Crystals',
Mineralogical Magazine Vol. 30, No. 220. London
1953.

Talbot Rice, D. *The Background of Art*. London: Nelson
and Son 1959.

Thompson, D'Arcy *On Growth and Form*. London and
New York: Cambridge University Press 1969.

Waelder, R. *Psychoanalytic Avenues to Art*. London:
Hogarth Press 1959.

Waldberg, P. *Surrealism*. London: Thames and Hudson
1966.

Watson, Baker *The World Beneath the Microscope*.
New Vision Studio 1955.

Worringer, W. *Abstraction and Empathy*. London:
Routledge and Kegan Paul 1953.